CONTENTS

ANCIENT NUMBERS

A long time ago, people counted using their fingers. Symbols were invented to show the numbers. The symbols are known as **numerals**.

The Egyptians used this system.

I	II	III	IIII	III II	III III	IIII III	IIII IIII	IIIII IIII	∧
1	2	3	4	5	6	7	8	9	10

∧∧∧ ∧∧	ℓ	ℓℓℓ ℓℓ	🪷
50	100	500	1000

This system is probably based on pictures of fingers.

This number system can become quite complicated when the numbers become large. For example, look at the date 1066.

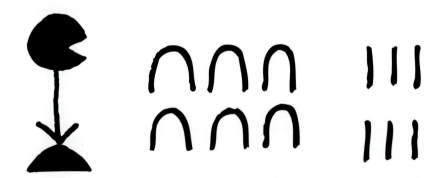

TO DO:

Write the numbers 125, 72 and 356 using the Egyptian system.

4

Numbers

David Kirkby

Heinemann

First published in Great Britain by Heinemann Library
an imprint of Heinemann Publishers (Oxford) Ltd
Halley Court, Jordan Hill, Oxford OX2 8EJ

MADRID ATHENS PARIS
FLORENCE PRAGUE WARSAW
PORTSMOUTH NH CHICAGO SAO PAULO
SINGAPORE TOKYO MELBOURNE AUCKLAND
IBADAN GABORONE JOHANNESBURG

Designed by The Point
Cover design by Pinpoint Design
Printed in China
Produced by Mandarin Offset
99 98 97 96 95
10 9 8 7 6 5 4 3 2 1

ISBN 0431 06897 6

British Library Cataloguing in Publication Data
Kirkby, David
Numbers. - (Maths Live Series)
I. Title II. Series
513.2

Acknowledgements
The author and publisher wish to acknowledge, with thanks,
the following photographic sources:
Trevor Clifford pp12, 37; all other photos: The Point.
The publishers would also like to thank the following for the kind loan of equipment:
NES Arnold Ltd; Polydron International Ltd.

Note to reader: words in **bold** in the text are explained in the glossary on page 44.

The Romans used several different numerals which can still be seen in lots of places today.

I	II	III	IV	V	VI	VII	VIII	IX	X
1	2	3	4	5	6	7	8	9	10

	L		C		D		M	
	50		100		500		1000	

Looking at the fingers on one hand, you can see what their symbol for 5 may have meant.
How do you think they arrived at the symbol for 10?

Notice how the Romans found short ways of writing some of their numbers. They wrote:

- the number 4 as IV to show that it is 'one less than five'
- the number 19 as XIX to show that it is 'one less than twenty'
- the number 90 as XC to show that it is 'ten less than one hundred', and so on.

Sometimes their large numbers are very difficult to read. For example, the year 1786 would be written:

MDCCLXXXVI

You read the number from left to right like this:

M	D	CC	L	XXX	VI	
1000	500	200	50	30	6	= 1786

CHALLENGE:

- What are these numbers?

    LXIII   CXIV   DCLIX   CMDL

- Write, using Roman numerals, some of your statistics. Include your age, your house number and your date of birth.

- Another problem with the Roman system is that calculations are very difficult. Try these:

$$\begin{array}{r} IV \\ +XI \\ \hline \\ \hline \end{array} \qquad \begin{array}{r} XIV \\ -VI \\ \hline \\ \hline \end{array}$$

2 WRITING NUMBERS

Nearly 1200 years ago, a famous Arab mathematician, Al Khowarizmi, wrote about an Indian way of writing numbers which he had discovered on his travels. His books reached continental Europe, where many Arabs lived.

Sometime later, about 800 years ago, a monk called Abelard, from Bath, travelled to Europe, translated the books of Al Khowarizmi, and brought them back to England. Gradually, our way of writing numbers changed.

In fact the numerals we use today have developed from Indian numerals first used about 1800 years ago.

Indian, 2nd century	— = ☰ Ƴ ɲ ℓ ƍ ɤ ろ
Indian, 9th century	১ ২ ৩ ৪ ৫ ২ ৭ ৲ ৭ ০
Arabic, 10th century	١ ٢ ₹ ৲ ৎ ৬ ৲ 8 9 ○
European, 12th century	١ ৎ ३ ३ Ƴ Ɫ ٦ 8 9
European, 15th century	١ ٢ ३ ৫ ٦ 6 ∧ 8 9 ○
European, 16th century	1 2 3 4 5 6 7 8 9 ◆

Since the way we write numbers comes from the Indians, then the Arabs, we call it the Hindu-Arabic system.
In this system, we only have to use these ten numerals:
   0, 1, 2, 3, 4, 5, 6, 7, 8 and 9.

These numerals are often known as **digits**. 'Digit' means 'finger', so using the word 'digit' for part of a number probably comes from the fact that we use fingers to count with.

The number 57 is a two-digit number, using the two digits 5 and 7. The number 481 is a three-digit number, using the digits 4, 8 and 1.

When we write a number, like 356, the position of each digit shows its value:

- the position of the 3 means three hundreds
- the position of the 5 means five tens
- the position of the 6 means six ones or six units.

So the positions of the digits in a three-digit number show the number of hundreds, the number of tens and the number of units.

Hundreds	Tens	Units
3	5	6

For a four-digit number, the positions of the digits show thousands, hundreds, tens and units.

The value of each position is ten times the value of the position to its right.

Zero is a very important digit in our number system. In the number 408, for example, the zero shows that there are no tens, and the number does not become confused with 48.

 CHALLENGE:

- Make four numbered cards: 2, 3, 6, 8.
- Put any two cards together to make a two-digit number.
- It is possible to make 12 different two-digit numbers. Can you find them?
- How many different three-digit numbers can you make, by choosing three cards from the four?

ORDERING NUMBERS

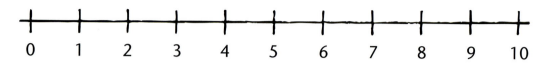

A **number line** is a way of showing the order of numbers.

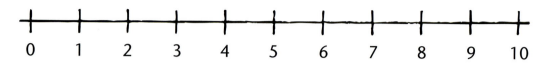

On this number line, the numbers become larger as you move from left to right.

CHALLENGE:

Order the cards

You need a pack of playing cards.

- Take out a set of ten cards numbered from 1 to 10.
- Shuffle your cards, then spread them out in a straight line.

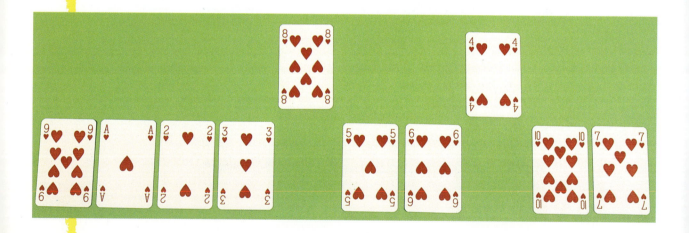

- You may swap any two cards at a time. Keep swapping until you have all the cards in the correct order.
- Count the number of swaps you needed.
- Try the activity a few more times, and see what is the fewest number of swaps you need.

A number line can start and finish at any number.

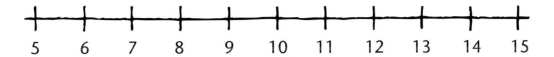

5   6   7   8   9   10   11   12   13   14   15

The number 8 is less than 12, which in mathematical shorthand is written as 8 < 12. The number 14 is greater than 10, which in mathematical shorthand is written as 14 > 10.

So < means 'is less than' and > means 'is greater than'.

The number 9 is between 7 and 11, which in mathematical shorthand is written as 7 < 9 < 11. This reads '7 is less than 9, which is less than 11', or '9 is between 7 and 11'.

TO DO:

Play the between game

You need three players and two dice.

- Take turns to throw the dice. Add up the total number of spots each player throws.
- The winner of the round is the player whose total is between the other two. If there is a tie, throw again.
- Play ten rounds, keep score and see who wins the most rounds.

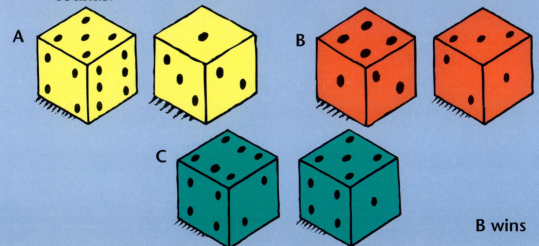

A

B

C

B wins

4 CALCULATOR NUMERALS

Calculator numerals do not look exactly like our written numerals. This is because they are based on a set of seven rods of equal lengths.

In a calculator display, the rods are shown by bars which light up. By lighting up some bars and not others, different numerals can be shown.

The same way of showing numerals is used in the display of a digital clock.

TO DO:

• Draw the time now, using coloured bars.

We can show the numeral 3 by lighting up five of the seven bars. *Which numeral needs the most bars lit? Which needs the fewest?* There are three numerals which need five bars lit. *Which are they?*

 ## CHALLENGE:

This calculator shows the two-digit number 44. It has a total of 8 lit bars. There are eight other two-digit numbers which need 8 lit bars. Can you find them?

5 ADDING

When you find the total number of objects which are in different groups, it is called **adding**. You add them together. The process of adding is called **addition**.

Annabel has 3 apples, and Paul has 5. Putting them together, there is a total of 8. When 3 and 5 are added, the total is 8. We write $3 + 5 = 8$, which reads '3 and 5 makes 8'.

The '+' sign is called a 'plus' sign. The '=' sign is called an 'equals' sign.

So, sometimes $3 + 5 = 8$ is read '3 plus 5 equals 8'. This addition can also be written like this:

$$\begin{array}{r} 3 \\ + 5 \\ \hline 8 \end{array}$$

An **addition table** shows the results of adding two numbers together. The entry in the table is the total of the row heading and column heading.

For example $4 + 3 = 7$

	1	2	3	4	5	6
1	2	3	4	5	6	7
2	3	4	5	6	7	8
3	4	5	6	7	8	9
4	5	6	7	8	9	10
5	6	7	8	9	10	11
6	7	8	9	10	11	12

TO DO:

Make your own addition table with the numbers 6, 7, 8, 9, 10 and 11 as headings.

When you order a meal, you need to know how much it will cost altogether. So you need to add the prices for all the items you order.

What will it cost you for a cheese and tomato pizza slice, a salad and a drink of Cola? Choose your own meal and find its cost.

TO DO:

Play the fifteens game

This is a game for two players. You need a set of nine cards, numbered from 1 to 9. Make a 3 by 3 square board, so that the cards will fit on to each small square.

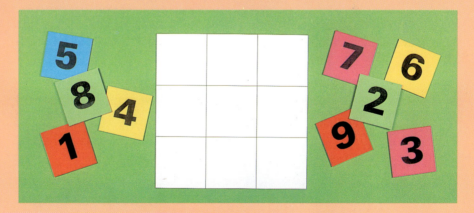

- One player takes the even-numbered cards; the other player takes the odd-numbered cards.
- 'Odd' plays first, by placing a card on one of the squares.
- Take turns placing a card.
- The winner is the first player to complete a line of three cards which total 15.

6 SUBTRACTING

If there is a collection of objects, from which you take some away, then count how many are left, you are **subtracting**. The process of subtracting is called **subtraction**.

There were 7 cakes on the plate and 2 are being taken away, which leaves 5 cakes. When 2 is subtracted from 7, the result is 5. We write 7 – 2 = 5, which reads '7 take away 2 leaves 5'.

The '–' sign is called a 'minus' sign.
The '=' sign is called an 'equals' sign.
So sometimes 7 – 2 = 5 is read '7 minus 2 equals 5'.
This subtraction can also be written like this:

$$\begin{array}{r} 7 \\ -\ 2 \\ \hline 5 \\ \hline \end{array}$$

We are also subtracting when we try to find the difference between two things.

Bejal is 120 centimetres tall.
Claire is 108 centimetres tall.
What is the difference between their heights?

In a game of darts, each player starts with a total of 301 points. On your turn, you throw three darts and subtract your score from your total.

This is Lynn's first throw.
How many points has she scored with her three darts? What should she write for her score on the board? What was Dean's score on his first throw?

Dean	Lynn
301	301
270	

TO DO:

Play the subtraction game

This is a game for several players. You need a pack of playing cards without the picture cards.

- Shuffle the cards and deal three each.
- Each player places their cards in this arrangement.
- The winner is the player with the largest answer to their subtraction.

You could play some variations of this game. The winner is the player with:

- the smallest answer
- the largest odd-numbered answer
- the smallest even-numbered answer
- the nearest answer to 40
- the nearest answer to 55.

CHALLENGE:

You need four cards numbered 2, 3, 5, 8.

- Arrange the cards to make 2 two-digit numbers, then subtract the smallest from the largest. What is the answer?
- It is possible to arrange the four cards here to give twelve different answers. The smallest is 2, and the largest is 62. Try this with your selection of cards.

7 MULTIPLYING

Multiplying is a quick way of adding. The process of multiplying is called **multiplication**.

There are 6 pieces of chocolate in each row. There are 3 rows. So altogether there are 3 lots of 6 pieces, which is 18 pieces. We say that '6 multiplied by 3 is 18', or '6 times 3 = 18', and we write 6 x 3 = 18. This multiplication can also be written like this:

$$\begin{array}{r} 6 \\ \times\ 3 \\ \hline 18 \end{array}$$

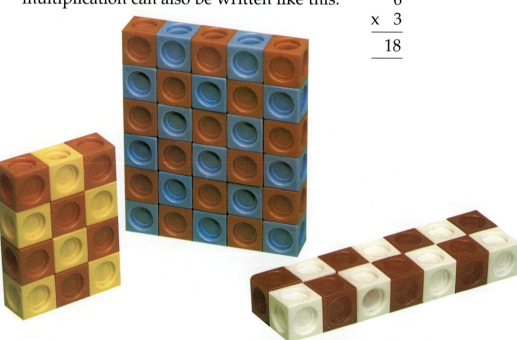

The set of cubes on the left shows that 3 multiplied by 4 is 12, or 3 x 4 = 12. *What multiplications do the other two sets of cubes show?*

The first set of stamps shows that 5 multiplied by 3 is 15. The other shows that 3 multiplied by 5 is 15. So 3 x 5 = 15, and 5 x 3 = 15.

The result of a multiplication is called the **product**. So we say that 'the product of 3 and 5 is 15'.

What multiplications do these bottles show?

CHALLENGE:

Make a multiplication square

- Copy the large square on the right.
- Choose any small square.
- Imagine that it is a corner of a rectangle, and that the other corner is in the top left of the large square.

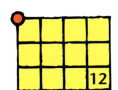 

- Count the number of squares inside the rectangle. Write it in the small square.
- Continue until you have filled all the small squares.

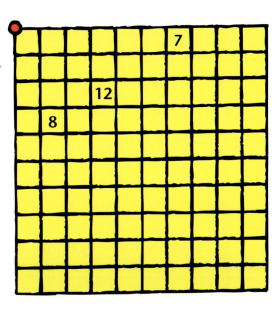

8 DIVIDING

When things are being shared out, it is called **dividing**.
The process of dividing is called **division**.

This picture shows how 15 biscuits have been shared out
equally among 3 plates. Each person's share is 5 biscuits.
So 15 has been divided by 3, to give 5 each.
We write 15 ÷ 3 = 5, which reads '15 divided by 3 is 5'.
We say that '15 is exactly divisible by 3'.

Division is very closely linked to multiplication.

How many marbles can you see?
The marbles are in 5 groups, with 4 in each group.

The total number of marbles is found by multiplying
5 by 4: 5 x 4 = 20
20 marbles divided by 5 groups tells you how many in
each group: 20 ÷ 5 = 4
20 marbles divided by 4 in each group tells you the
number of groups: 20 ÷ 4 = 5
So for every multiplication fact, there are two linked
division facts.

1	2	3	4	5	6	7	8	9	10
2	4	6	8	10	12	14	16	18	20
3	6	9	12	15	18	21	24	27	30
4	8	12	16	20	24	28	32	36	40
5	10	15	20	25	30	35	40	45	50
6	12	18	24	30	36	42	48	54	60
7	14	21	28	35	42	49	56	63	70
8	16	24	32	40	48	56	64	72	80
9	18	27	36	45	54	63	72	81	90
10	20	30	40	50	60	70	80	90	100

Every multiplication fact in the multiplication square leads to two linked division facts. The highlighted square, for example, shows the multiplication fact, 5 x 7 = 35. From this we know that 35 ÷ 5 = 7, and 35 ÷ 7 = 5.

TO DO:

- Use the multiplication square to find ten numbers which are exactly divisible by 6.
- Write down the 20 division facts.

Sometimes things divide exactly, and sometimes they do not. If you are sharing some toffees, for example, and they do not divide exactly, then you may have one or two toffees left over. This is called a **remainder**.

In this picture, 14 toffees have been divided by 4 to give 3 each, and there is a remainder of 2. We write 14 ÷ 4 = 3 remainder 2.

CHALLENGE:

Which numbers in the multiplication square will leave a remainder of 1 when divided by 5?

MULTIPLICATION TABLES

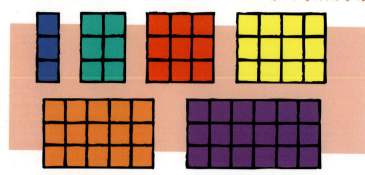

Each of these sets of squares shows a different multiplication by 3. They are:

1 times 3 is 3	or	1 x 3 = 3
2 times 3 is 6	or	2 x 3 = 6
3 times 3 is 9	or	3 x 3 = 9
4 times 3 is 12	or	4 x 3 = 12
5 times 3 is 15	or	5 x 3 = 15
6 times 3 is 18	or	6 x 3 = 18

This is called a **multiplication table** for multiplying by 3. It is sometimes called a times table for multiplying by 3, or a 3 times table.

The factbox gives some more multiplication tables.

FACTBOX

1 x 8 = 8	1 x 3 = 3	1 x 5 = 5	1 x 7 = 7
2 x 8 = 16	2 x 3 = 6	2 x 5 = 10	2 x 7 = 14
3 x 8 = 24	3 x 3 = 9	3 x 5 = 15	3 x 7 = 21
4 x 8 = 32	4 x 3 = 12	4 x 5 = 20	4 x 7 = 28
5 x 8 = 40	5 x 3 = 15	5 x 5 = 25	5 x 7 = 35
6 x 8 = 48	6 x 3 = 18	6 x 5 = 30	6 x 7 = 42
7 x 8 = 56	7 x 3 = 21	7 x 5 = 35	7 x 7 = 49
8 x 8 = 64	8 x 3 = 24	8 x 5 = 40	8 x 7 = 56
9 x 8 = 72	9 x 3 = 27	9 x 5 = 45	9 x 7 = 63
10 x 8 = 80	10 x 3 = 30	10 x 5 = 50	10 x 7 = 70

CHALLENGE:

Here are the next five lines of the 3 times table. What are the next five lines of the other tables?

11 x 3 = 33
12 x 3 = 36
13 x 3 = 39
14 x 3 = 42
15 x 3 = 45

1	2	3	4	5	6	7	8	9	10
2	4	6	8	10	12	14	16	18	20
3	6	9	12	15	18	21	24	27	30
4	8	12	16	20	24	28	32	36	40
5	10	15	20	25	30	35	40	45	50
6	12	18	24	30	36	42	48	54	60
7	14	21	28	35	42	49	56	63	70
8	16	24	32	40	48	56	64	72	80
9	18	27	36	45	54	63	72	81	90
10	20	30	40	50	60	70	80	90	100

The rows and columns of a multiplication square show all the multiplication tables for multiplying by 1 up to multiplying by 10. If you look down the fourth column, you see the results of multiplying by 4.

Look carefully at the multiplication square, and you will see that some numbers appear several times. For example, the number 12 appears four times, and the number 16 appears three times.

CHALLENGE:

- Apart from number 12, there are seven more numbers which appear four times. Can you spot them?
- Apart from number 16, there are three more numbers which appear three times. Can you spot these?
- There are nineteen numbers up to 50 which do not appear at all. They are, in order:

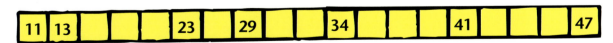

| 11 | 13 | | | | 23 | | 29 | | 34 | | | | 41 | | | | 47 |

Can you fill in the gaps?

10 MULTIPLES

The **multiples** of 3 are the results of multiplying different numbers by 3. So the multiples of 3 are 3, 6, 9, 12, 15 and so on. To show the multiples of 3, draw a 3-row grid and write the numbers in order, starting at 1, like this:

1	4	7	10	13	16	19	22	25	28
2	5	8	11	14	17	20	23	26	29
3	6	9	12	15	18	21	24	27	30

FACTBOX
1 x 3 = **3**
2 x 3 = **6**
3 x 3 = **9**
4 x 3 = **12**
5 x 3 = **15**
6 x 3 = **18**
7 x 3 = **21**
8 x 3 = **24**
9 x 3 = **27**
10 x 3 = **30**

The multiples of 3 will appear in the bottom row.

TO DO:

Use the same method to show the multiples of 7.
You will need to draw a 7-row grid on squared paper.

The multiples of 3 make an interesting pattern on a hundred square. *How would you describe the pattern?*

1	2	3	4	5	6	7	8	9	10
11	12	13	14	15	16	17	18	19	20
21	22	23	24	25	26	27	28	29	30
31	32	33	34	35	36	37	38	39	40
41	42	43	44	45	46	47	48	49	50
51	52	53	54	55	56	57	58	59	60
61	62	63	64	65	66	67	68	69	70
71	72	73	74	75	76	77	78	79	80
81	82	83	84	85	86	87	88	89	90
91	92	93	94	95	96	97	98	99	100

The multiples of other numbers are found by looking along the rows, or down the columns of a multiplication square. So the multiples of 7, for example, are: 7, 14, 21, 28, 35 and so on.

Notice that 15 is a multiple of 5, and also that 15 is a multiple of 3. So 15 is a common multiple of 3 and 5.

18 is a multiple of 6, and 18 is also a multiple of 9. So 18 is a common multiple of 6 and 9.

1	2	3	4	5	6	7	8	9	10
2	4	6	8	10	12	14	16	18	20
3	6	9	12	15	18	21	24	27	30
4	8	12	16	20	24	28	32	36	40
5	10	15	20	25	30	35	40	45	50
6	12	18	24	30	36	42	48	54	60
7	14	21	28	35	42	49	56	63	70
8	16	24	32	40	48	56	64	72	80
9	18	27	36	45	54	63	72	81	90
10	20	30	40	50	60	70	80	90	100

CHALLENGE:

Find two numbers in the multiplication square which are common multiples of 4 and 5.

TO DO:

Play 'Up the multiples of 3, down the multiples of 5'

This is a game for two players.

START	1	2	3	4	5	6	7	8	9	10	11	12	13
28	27	26	25	24	23	22	21	20	19	18	17	16	15	14
29	30	31	32	33	34	35	36	37	38	39	40	41	42	43
58	57	56	55	54	53	52	51	50	49	48	47	46	45	44
59	60	61	62	63	64	65	66	67	68	69	70	FINISH

- Place a counter each at 'Start'.
- Take turns to throw a dice, and move that number of spaces round the board.
- If you land on a multiple of 3, you jump forwards to the next multiple of 3.
- If you land on a multiple of 5, you jump backwards to the previous multiple of 5.
- The winner is the first to reach the 'Finish'.

Play again, choosing a different multiple for 'up', and another for 'down'.

FACTORS

The number 18 can be divided exactly by six numbers, 1, 2, 3, 6, 9 and 18. The numbers 1, 2, 3, 6, 9 and 18 are called the **factors** of 18. The number 18 has six factors.

The factors of a number are all the numbers which will divide into it exactly. As 3 is a factor of 18, it follows that 18 is a multiple of 3.

12 flower pots have been arranged in different ways. The picture shows them arranged in:

- 2 rows of 6 pots, which shows that 2 and 6 are a pair of factors of 12
- 3 rows of 4 pots, which shows that 3 and 4 are a pair of factors of 12
- 1 row of 12 pots, which shows that 1 and 12 are a pair of factors of 12

So 12 has three pairs of factors.

TO DO:

You need some counters to stand for your flower pots.

- Start with 20 counters.
- Find different ways of arranging them in equal rows.
- Write down the pair of factors shown by each arrangement.
- Finally, list all the factors of 20.

Try again, with a different number of counters.

FACTBOX

Number	Factors	Number	Factors
1	1	16	1, 2, 4, 8, 16
2	1, 2	17	1, 17
3	1, 3	18	1, 2, 3, 6, 9, 18
4	1, 2, 4	19	1, 19
5	1, 5	20	1, 2, 4, 5, 10, 20
6	1, 2, 3, 6	21	1, 3, 7, 21
7	1, 7	22	1, 2, 11, 22
8	1, 2, 4, 8	23	1, 23
9	1, 3, 9	24	1, 2, 3, 4, 6, 8, 12, 24
10	1, 2, 5, 10	25	1, 5, 25
11	1, 11	26	1, 2, 13, 26
12	1, 2, 3, 4, 6, 12	27	1, 3, 9, 27
13	1, 13	28	1, 2, 4, 7, 14, 28
14	1, 2, 7, 14	29	1, 29
15	1, 3, 5, 15	30	1, 2, 3, 5, 6, 10, 15, 30

Notice that 1 is a factor of every number. Notice, too, that every number is a factor of itself. For example, 10 is a factor of 10.

CHALLENGE:

The number 3 is a factor of 6 and a factor of 21. So 3 is a common factor of 6 and 21. Can you find some common factors of 20 and 24, 8 and 18, 15 and 10, 16 and 30?

TO DO:

Factor Game Scoresheet

☐	☐	has a factor of 2
☐	☐	has a factor of 3
☐	☐	has a factor of 4
☐	☐	has a factor of 5
☐	☐	has a factor of 6
☐	☐	has a factor of 7
☐	☐	has a factor of 8
☐	☐	has a factor of 9

Play the factor game

- Shuffle a set of cards numbered from 1 to 9.
- Each player makes a scoresheet like the one shown.
- Deal out four of the cards, such as:

- Using these four digits only, and not using any one of them more than four times altogether, see how many two-digit numbers you can complete correctly on the scoresheet.

PRIME NUMBERS

Prime numbers are numbers which have exactly two factors. For example, the only factors of 7 are 1 and 7; the only factors of 11 are 1 and 11. So 7 and 11 are prime numbers.

7	17	**13**
11	37	**5**

Prime numbers can be found by using the **Sieve of Eratosthenes**. Eratosthenes, a Greek mathematician who lived in about 200 BC, devised the following method for finding prime numbers.

TO DO:

You need counters of five different colours.

- Start with a hundred square.
- Place one coloured counter on the number 1.
- Place counters of the next colour on the multiples of 2, except 2 itself.
- Place counters of the next colour on the multiples of 3, except 3 itself.
- Place counters of the next colour on the multiples of 5, except 5 itself.
- Finally, place counters of the last colour on the multiples of 7, except 7 itself.

1	2	3	4	5	6	7	8	9	10
11	12	13	14	15	16	17	18	19	20
21	22	23	24	25	26	27	28	29	30
31	32	33	34	35	36	37	38	39	40
41	42	43	44	45	46	47	48	49	50
51	52	53	54	55	56	57	58	59	60
61	62	63	64	65	66	67	68	69	70
71	72	73	74	75	76	77	78	79	80
81	82	83	84	85	86	87	88	89	90
91	92	93	94	95	96	97	98	99	100

The numbers which have counters on them have been sifted out. The numbers which are left without a counter are the prime numbers up to 100. How many are there?

Numbers which are not prime numbers can be arranged to form a square or rectangle.

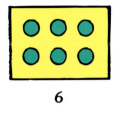

6

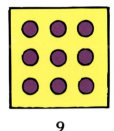

9

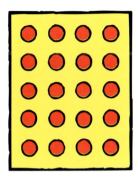

20

These numbers have more than two factors. Prime numbers cannot be arranged to form a square or rectangle.

If the numbers from 1 to 100 are written, in order, in six columns, and then the prime numbers are highlighted, an interesting pattern can be seen.

This shows that, apart from the first two prime numbers, 2 and 3, they all lie in either the first or fifth column.

In other words, they are all either one more or one less than a multiple of 6.

Notice that 2 is the only even prime number. Notice, too, that all prime numbers have 1, 3, 7 or 9 as their last digit.

1	2	3	4	5	6
7	8	9	10	11	12
13	14	15	16	17	18
19	20	21	22	23	24
25	26	27	28	29	30
31	32	33	34	35	36
37	38	39	40	41	42
43	44	45	46	47	48
49	50	51	52	53	54
55	56	57	58	59	60
61	62	63	64	65	66
67	68	69	70	71	72
73	74	75	76	77	78
79	80	81	82	83	84
85	86	87	88	89	90
91	92	93	94	95	96
97	98	99	100		

CHALLENGE:

- Can you write the first ten prime numbers without looking at this book?
- Now can you write the next five?

27

13 SQUARE NUMBERS

If a number of pegs can be placed on a pegboard to make a square, then the number of pegs used is called a **square number**.

 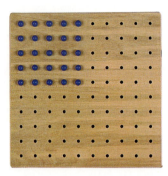 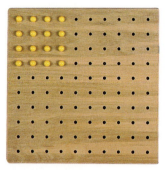

These three pegboards show 9 pegs, 25 pegs and 16 pegs, in turn, arranged in squares. They show that 9, 25 and 16 are square numbers.

The square numbers are the results of the multiplication facts: 1 x 1, 2 x 2, 3 x 3, 4 x 4 and so on.
They are 1, 4, 9, 16, 25, 36, 49, 64, 81, 100 and so on.

We write:
- $2^2 = 4$, and say 'two squared is four'
- $3^2 = 9$, and say 'three squared is nine'
- $4^2 = 16$, and say 'four squared is sixteen'
- $5^2 = 25$, and say 'five squared is twenty-five'.

1	2	3	4	5	6	7	8	9	10
2	4	6	8	10	12	14	16	18	20
3	6	9	12	15	18	21	24	27	30
4	8	12	16	20	24	28	32	36	40
5	10	15	20	25	30	35	40	45	50
6	12	18	24	30	36	42	48	54	60
7	14	21	28	35	42	49	56	63	70
8	16	24	32	40	48	56	64	72	80
9	18	27	36	45	54	63	72	81	90
10	20	30	40	50	60	70	80	90	100

Note their position on the multiplication square. They lie along a diagonal line.

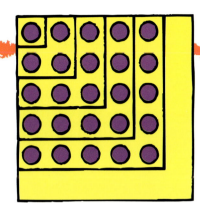

This picture shows a set of different-sized squares, each with a corner in the top left. Look at the picture so that you can see the different square numbers.

TO DO:

- Notice that you can build up one square number from another by adding on some pegs.

  Write the square numbers

  1   4   9   16   25   36   49 …

  Write the differences

   (+3)   (+5)   (+7)   (+9)   (+11)   (+13) …

- Notice also that $4 = 1 + 3$
  $9 = 1 + 3 + 5$
  $16 = 1 + 3 + 5 + 7$

Write the next three square numbers like this.
Can you see another pattern?

TO DO:

- Make your own number spiral like this on squared paper, then colour all the square numbers.
- What pattern do the square numbers make?

	26	27	28	29	30	31		
	25	10	11	12	13	32		
	24	9	2	3	14	33		
	23	8	1	4	15	34		
	22	7	6	5	16	35		
	21	20	19	18	17	36		
						37		

14 FRACTIONS

A **fraction** is a part of a whole. For example, if you slice a pizza into two equal pieces, then each piece is called one half. If, instead, you slice it into three equal pieces, each is called one third. If four people are going to share the pizza, then you need four equal pieces, and each piece is called one quarter. One half, one third and one quarter are all examples of fractions.

The fraction one half is written as $\frac{1}{2}$.
One third is written $\frac{1}{3}$ and one quarter is written $\frac{1}{4}$.

Other examples of fractions are shown in the factbox.

FACTBOX		
Number of equal parts of a whole	**Fraction of each part**	
2	one half	$\frac{1}{2}$
3	one third	$\frac{1}{3}$
4	one quarter	$\frac{1}{4}$
5	one fifth	$\frac{1}{5}$
6	one sixth	$\frac{1}{6}$
7	one seventh	$\frac{1}{7}$
8	one eighth	$\frac{1}{8}$
9	one ninth	$\frac{1}{9}$
10	one tenth	$\frac{1}{10}$

TO DO:

Find different halves

You need some squared paper.

- Draw some 4 by 4 squares.
- Here are three ways of halving the square.
- How many other ways can you find?

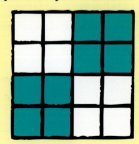
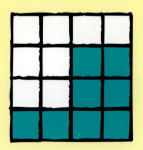

The cake has been divided into three thirds. There is one third on the right, and two thirds on the left. We write these fractions as $\frac{1}{3}$ and $\frac{2}{3}$.

Fractions of each shape have been coloured.
The fractions are three quarters of the square, two fifths of the pentagon, five sixths of the hexagon and seven eighths of the rectangle.
Can you say what fraction of each shape is not coloured?

 CHALLENGE:

Each of these squares has been divided into parts.
The lines drawn to the sides of each square meet the sides half-way. Can you say what fraction each part is of each square? One is done for you.

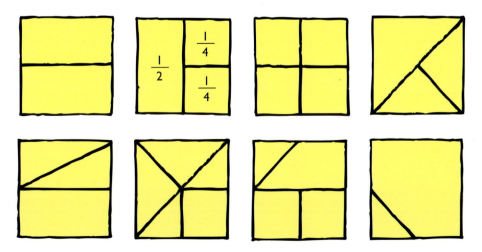

31

15 EQUIVALENT FRACTIONS

Different fractions can sometimes mean the same part of a whole.

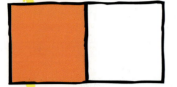  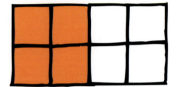

For example, the drawing shows that one half is the same as two quarters, which is the same as four eighths of the rectangle.

So $\frac{1}{2}$, $\frac{2}{4}$ and $\frac{4}{8}$ are called **equivalent fractions**.

One way of showing equivalent fractions is to build a fraction wall. Each layer of the wall is one whole, and this is marked on the bottom layer. The other layers are divided into equal parts, and their fractions are marked on the wall.

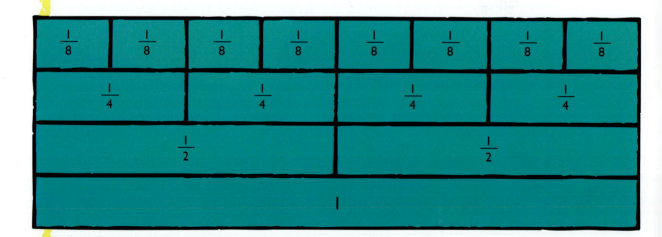

From the wall you can see that $\frac{1}{2}$ is equivalent to $\frac{2}{4}$, which is equivalent to $\frac{4}{8}$.

$\frac{1}{4}$ and $\frac{2}{8}$ are also equivalent fractions.

So are $\frac{3}{4}$ and $\frac{6}{8}$.

1	2	3	4	5	6	7	8	9	10
2	4	6	8	10	12	14	16	18	20
3	6	9	12	15	18	21	24	27	30
4	8	12	16	20	24	28	32	36	40
5	10	15	20	25	30	35	40	45	50
6	12	18	24	30	36	42	48	54	60
7	14	21	28	35	42	49	56	63	70
8	16	24	32	40	48	56	64	72	80
9	18	27	36	45	54	63	72	81	90
10	20	30	40	50	60	70	80	90	100

A good way of finding sets of equivalent fractions is to use a multiplication square.

To find some fractions equivalent to $\frac{2}{5}$, look along the 2 row, and also along the 5 row.

Pairing the numbers, one from each row, gives this set of equivalent fractions: $\frac{2}{5} = \frac{4}{10} = \frac{6}{15} = \frac{8}{20} = \frac{10}{25}$ and so on.

CHALLENGE:

Use the multiplication square to find some fractions equivalent to $\frac{3}{4}$ and also to $\frac{5}{6}$.

TO DO:

Make some equivalent fractions

- Take a piece of paper.
- Fold it in half, then in half again, and finally once more, so that the folds are as shown above.
- Open out the paper and mark each section with its fraction of the whole piece of paper.
- Can you use this paper to show some equivalent fractions?

Experiment by folding other pieces of paper in different ways, looking for equivalent fractions.

16 DECIMALS

Our way of writing numbers is based on the position of the digits 0, 1, 2, 3, 4, 5, 6, 7, 8 and 9. So, for example, 273 means 2 hundreds, 7 tens and 3 ones, and is read 'Two hundred and seventy-three'.

Thousands	Hundreds	Tens	Ones
	2	7	3

The value of each position is ten times the value of the position to the right. So if we continue the order:

hundreds, tens, ones …

then the value of the next position is tenths.

To separate the whole numbers from the fraction, a point is placed between them. This is known as a **decimal point**, and a number which has a decimal point is called a **decimal number**. For example, 358.6 means 3 hundreds, 5 tens, 8 ones and 6 tenths, and is read 'three hundred and fifty-eight point six'.

Thousands	Hundreds	Tens	Ones	Tenths
	3	5	8 •	6

Sometimes a number has no whole numbers, just tenths. 0.3 means 0 ones and 3 tenths, and is read 'nought point three'.

CHALLENGE:

You need four numbered cards.

- On a large piece of paper, draw boxes for a decimal number like this:

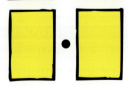

- Choose two cards to make a decimal number with ones and tenths.
- It is possible to make 12 different decimal numbers. Can you find them?
- Put them in order, from smallest to largest.

Just as with whole numbers, we can show decimal numbers on number lines.

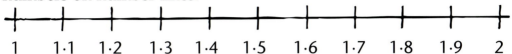

| 1 | 1·1 | 1·2 | 1·3 | 1·4 | 1·5 | 1·6 | 1·7 | 1·8 | 1·9 | 2 |

This number line shows the tenths between the whole numbers 1 and 2.

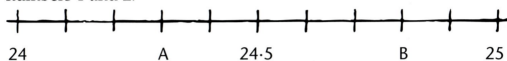

| 24 | | A | | 24·5 | | | B | | 25 |

This number line is not completely marked.
What numbers do A and B stand for, on the line?

The markings on a ruler show another number line marked in centimetres and millimetres. There are 10 millimetres in each centimetre, so each millimetre is one tenth of a centimetre.

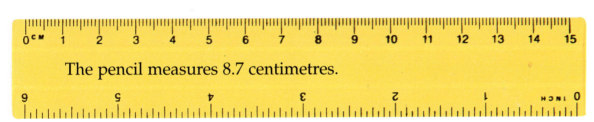

The pencil measures 8.7 centimetres.

TO DO:

Play the decimal guessing game

This is a game for two players

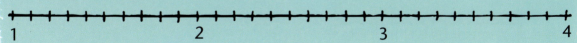

| 1 | 2 | 3 | 4 |

- One player thinks of a decimal point on this number line, without saying what it is.
- The other player has to guess the number by asking a series of questions, to which the answer can only be 'Yes' or 'No'.
- Count the number of questions needed to guess the number.
- Then swap roles and see who needs the fewer questions.

17 HUNDREDTHS

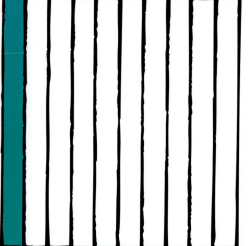 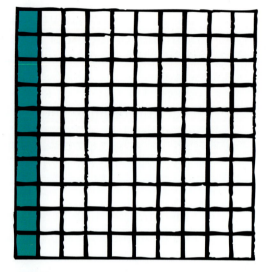

The square on the left has been divided into 10 equal parts. Each strip is one tenth of the square. The square on the right has been divided into 100 equal parts. Each small square is one hundredth of the large square. The coloured parts of each square show that ten hundredths is the same as one tenth.

We know that when numbers are written, the value of each position is ten times the value of the position to the right.

So if we continue the order:
hundreds, tens, ones, tenths . . .
then the value of the next position is hundredths.

For example, the decimal number 27·48 means 2 tens, 7 ones, 4 tenths and 8 hundredths, and is read 'twenty-seven point four eight'.

Thousands	Hundreds	Tens	Ones	Tenths	Hundredths
		2	7 •	4	8

Sprint races are measured in hundredths of a second.
These are some runners' times:

Smith   10·23 seconds   Jones   10·04 seconds
Brown   9·93 seconds   Collins   9·50 seconds
Wilson   9·86 seconds

Who won? What were the positions of the other athletes?

36

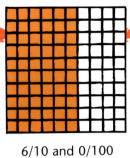

6/10 and 0/100
0·60

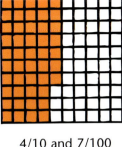

4/10 and 7/100
0·47

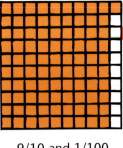

9/10 and 1/100
0·91

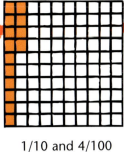

1/10 and 4/100
0·14

The first position after the decimal point shows tenths.
The second position after the decimal point shows
hundredths. These are sometimes called the first decimal
place and the second decimal place.
What do you think will be the value of the third decimal place?

TO DO:

Find some squared paper, and cut out four 10 by 10
squares.

• Colour fractions of each square to show the decimal
  numbers 0·35, 0·86, 0·7 and 0·4.

Height can be measured in metres and
centimetres. There are 100 centimetres in
each metre, so each centimetre is one
hundredth of a metre.

Siobhan's height is 112 centimetres.
So she is 1 metre and 12 centimetres tall.
Her height is 1·12 metres.

TO DO:

• Find a metric measure, and measure
  the heights of family and friends.
• Write their heights in metres using decimal numbers.

18 ROUNDING

Sometimes, instead of writing a number exactly, we round it. There are different ways of rounding a number.

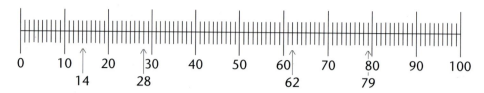

We can round up to the next 10, so that:
14 becomes 20   28 becomes 30   62 becomes 70
79 becomes 80   *What happens to 36 and 81?*

We can round down to the previous 10, so that:
14 becomes 10   28 becomes 20   62 becomes 60
79 becomes 70   *What happens to 33 and 58?*

We can round to the nearest 10, so that:
14 becomes 10   28 becomes 30   62 becomes 60
79 becomes 80   *What happens to 57 and 73?*

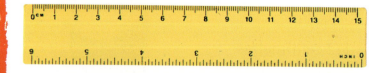

We sometimes round a measurement to the nearest unit.
What is the length of the carrot to the nearest centimetre?
What is the weight of the potatoes to the nearest kilogram?

TO DO:

• Measure your height, rounded to the nearest centimetre.
• Measure your weight, rounded to the nearest kilogram.

Sometimes, especially when numbers are large, we do not need to know the exact size of a number, but need a rough idea of its size. So we round the number. This is called approximating.

For example, if the number of people who watched a football match was 17, 472, we can approximate it to the nearest thousand: 17, 000. We can approximate it to the nearest hundred: 17, 500. Or we can approximate it to the nearest ten: 17, 470.

CHALLENGE:

- Can you round the attendance figures at each match to the nearest thousand?
- Now try it to the nearest hundred.

Tottenham Hotspur v West Ham	33,648
Chelsea v Manchester United	36,329
Everton v Leeds United	34,793
Arsenal v Sheffield Wednesday	26,522

TO DO:

You need a stopwatch on a digital watch.
- Set the stopwatch to zero.
  00:00 00
- Start it, then after a short while, stop it.
  00:16 79
- Write down the time on the watch.
- Write down the number of seconds which have passed.
- Round it to the nearest second.

Stopwatch time	Number of seconds	Nearest second
00:16 79	16·79	17

- Try some more times.

19 PERCENTAGES

A **percentage** is a fraction of 100. 100 per cent is the whole amount.

The fraction thirteen hundredths ($\frac{13}{100}$) is the percentage 13. We say it is 'thirteen per cent', and we write 13%. The fraction forty-seven hundredths ($\frac{47}{100}$) is the percentage 47. We say it is 'forty-seven per cent', and we write 47%.

In each of these squares, a different percentage has been coloured:

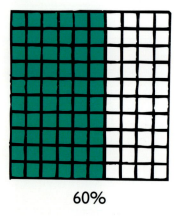

60%   32%

Well-known fractions of these squares have been coloured:

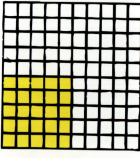 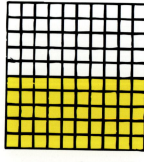 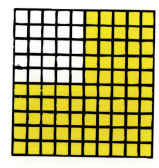

one quarter, $\frac{1}{4}$ is 25%   one half, $\frac{1}{2}$ is 50%   three quarters, $\frac{3}{4}$ is 75%

There are 10 sweets in this picture. So each sweet is 10% of the total (100%). 20% are green.
What percentages are yellow, red and pink?

> Holly pulled herself into a ball on the floor of the garage and waited for the shuddering to stop.
>
> 'Wake up!' Jamie ran back and crouched over her. 'Holly, wake up!'
>
> 'It's OK, look, she's breathing!' There was

Letters are either vowels – a, e, i, o, u – or consonants. When we write, do we use more vowels than consonants? Look at the passage of writing opposite. The two lines are drawn to mark off the first 100 letters. There are 38 vowels. Count them to check. There are 62 consonants. So, of these 100 letters, 38% are vowels, and 62% are consonants.

CHALLENGE:

- Find your own page of writing.
- Mark off the first 100 letters.
- Find the percentage of them which are vowels and the percentage which are consonants.
- How do they compare with the above percentages?

TO DO:

Which colours of car do you think are most popular?
- Record the colour of the first 100 cars you see.
- Start by making a chart to keep record.

Colour	Tally	Percentage
White	‖‖ ‖	
Black	‖‖	
Blue	‖‖ ‖‖	
Red	‖‖ ‖	
...		

- Stop when you have recorded the colours of 100 cars.
- Then write down the percentage of each colour.
- What are your conclusions?

POSITIVE AND NEGATIVE NUMBERS

A **number line** is used to show how numbers are ordered.

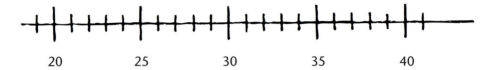

A number line is infinitely long, which means that it can be extended in both directions for ever. As we move along the line from left to right, the numbers get larger. As we move along the line from right to left, the numbers get smaller, and eventually reach 0.

As we count from right to left, the numbers are 6, 5, 4, 3, 2, 1 and then 0. Then, counting past 0, the next number is −1, then −2, then −3, and so on. We say these numbers as 'negative one', 'negative two', 'negative three', or as 'minus one', 'minus two' and 'minus three'.

The numbers below 0 are called **negative numbers**, while those above zero are called **positive numbers**.

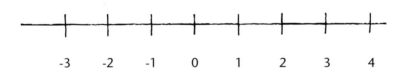

A thermometer is an instrument for measuring temperature. The scale on a thermometer is like a number line. Temperature is measured in degrees Celcius (written °C).

On a warm day, the temperature is about 20 °C, but in the winter, when the weather is much colder, the temperature comes down, sometimes as far as 0 °C, and sometimes even further.

At a temperature of 0 °C, water turns to ice. This temperature is called **freezing point**. Temperatures less than this are negative and below freezing point. A temperature of − 3 °C, for example, is termed '3 degrees below freezing'.

Positive and negative numbers can be added together.
For example, the result of adding −2 and 3 is 1,
while the result of adding 1 and −4 is −3. A simple way of
showing this is to jump along a number line.

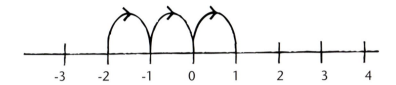

When you add a positive number, jump right.
When you add a negative number, jump left.

To add −2 and 3, start at the first number, −2, then jump 3
spaces right.
To add 1 and −4, start at the first number, 1, then jump 4
spaces left.

CHALLENGE:

• Make nine number cards like this:

| -4 | -3 | -2 | -1 | 0 | 1 | 2 | 3 | 4 |

• Make addition sums using three
  cards, like this:

• How many different additions
  can you make?

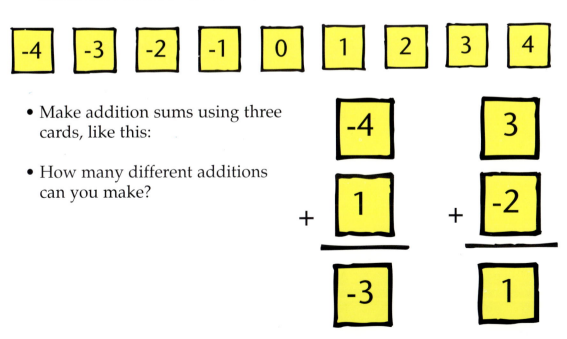

GLOSSARY

adding Finding the total number of objects in different groups.

addition The process of **adding**.

addition table A number table which allows you to see the result of **adding** two numbers together.

decimal number A number which is partly a whole number and partly a **fraction**.

decimal point A point used to separate the whole numbers from the **fraction** in a **decimal number**.

digit A **numeral** used as part of a number.

dividing Sharing a group of objects equally.

division The process of **dividing**.

equivalent fractions Different **fractions** which are the same part of a whole.

factor The factors of a number are those numbers which will divide into it exactly.

fraction A part of a whole.

freezing point The temperature at which water freezes.

multiples The results of a **multiplication table**.

multiplication The process of **multiplying**.

multiplication table A number table which allows you to see the result of **multiplying** two numbers together

multiplying **Adding** several groups of objects each of which contains the same number.

negative numbers Numbers below zero.

number line A straight line with equal divisions to show numbers in order.

numerals Symbols used to show a number.

percentage A number of hundredths of a whole.

positive numbers Numbers above zero.

prime number	A number which has exactly two factors.
product	The result of a **multiplication**.
remainder	The objects left when a **division** will not give exactly equal shares.
Sieve of Eratosthenes	A method for finding **prime numbers**.
square number	The result of **multiplying** a number by itself.
subtracting	Taking a number of objects away from a group and counting the number left.
subtraction	The process of **subtracting**.

INDEX

ANSWERS

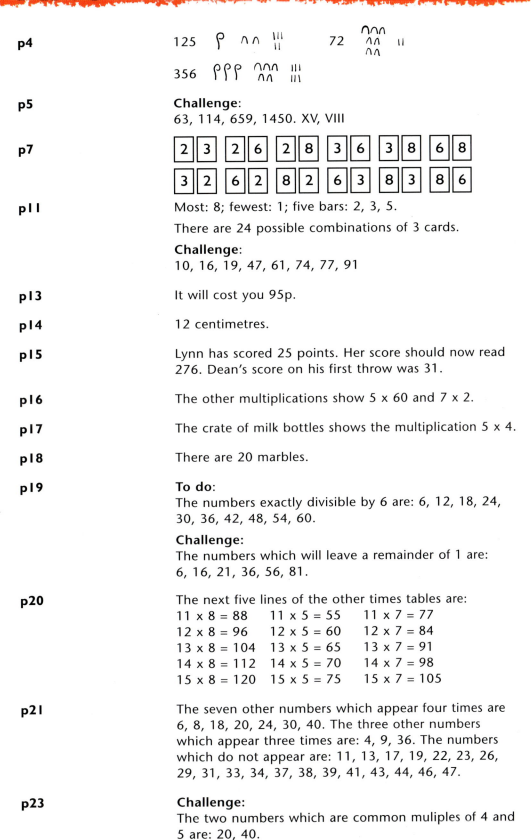

p4 125 𓊪 ∧∧ ||| 72 ∧∧∧ ∧∧ ∧∧ ||

356 𓊪𓊪𓊪 ∧∧∧ ∧∧ |||
|||

p5 **Challenge:**
63, 114, 659, 1450. XV, VIII

p7
2	3	2	6	2	8	3	6	3	8	6	8
3	2	6	2	8	2	6	3	8	3	8	6

p11 Most: 8; fewest: 1; five bars: 2, 3, 5.

There are 24 possible combinations of 3 cards.

Challenge:
10, 16, 19, 47, 61, 74, 77, 91

p13 It will cost you 95p.

p14 12 centimetres.

p15 Lynn has scored 25 points. Her score should now read 276. Dean's score on his first throw was 31.

p16 The other multiplications show 5 x 60 and 7 x 2.

p17 The crate of milk bottles shows the multiplication 5 x 4.

p18 There are 20 marbles.

p19 **To do:**
The numbers exactly divisible by 6 are: 6, 12, 18, 24, 30, 36, 42, 48, 54, 60.

Challenge:
The numbers which will leave a remainder of 1 are: 6, 16, 21, 36, 56, 81.

p20 The next five lines of the other times tables are:
11 x 8 = 88	11 x 5 = 55	11 x 7 = 77
12 x 8 = 96	12 x 5 = 60	12 x 7 = 84
13 x 8 = 104	13 x 5 = 65	13 x 7 = 91
14 x 8 = 112	14 x 5 = 70	14 x 7 = 98
15 x 8 = 120	15 x 5 = 75	15 x 7 = 105

p21 The seven other numbers which appear four times are 6, 8, 18, 20, 24, 30, 40. The three other numbers which appear three times are: 4, 9, 36. The numbers which do not appear are: 11, 13, 17, 19, 22, 23, 26, 29, 31, 33, 34, 37, 38, 39, 41, 43, 44, 46, 47.

p23 **Challenge:**
The two numbers which are common muliples of 4 and 5 are: 20, 40.

p24   The factors of 20 are: 1, 2, 4, 5, 10.

p25   **Challenge:**
Common factors of 20 and 24: 1, 2, 4
Common factors of 8 and 18: 1, 2
Common factors of 15 and 10: 1, 5
Common factors of 16 and 30: 1, 2

p26   **To Do:**
There are 25 prime numbers.

Challenge:
The first 10 prime numbers are: 1, 2, 3, 5, 7, 11, 13, 17, 19, 23. The next 5 are: 29, 31, 37, 41, 43

p31 $\frac{1}{4}, \frac{3}{5}, \frac{1}{6}, \frac{1}{8}$

Challenge:
The fractions are: $\frac{1}{2}, \frac{1}{2}$. $\frac{1}{2}, \frac{1}{4}, \frac{1}{4}$. $\frac{1}{4}, \frac{1}{4}, \frac{1}{4}, \frac{1}{4}$. $\frac{1}{2}, \frac{1}{4}, \frac{1}{4}$.

$\frac{1}{4}, \frac{1}{4}, \frac{1}{2}$. $\frac{1}{4}, \frac{1}{4}, \frac{1}{4}, \frac{1}{8}, \frac{1}{8}$. $\frac{1}{8}, \frac{3}{8}, \frac{1}{4}, \frac{1}{4}$. $\frac{1}{8}, \frac{7}{8}$.

p33   **Challenge:**
The equivalent fractions of $\frac{3}{4}$ are: $\frac{6}{8}, \frac{9}{12}, \frac{12}{16}, \frac{15}{20}, \frac{18}{24}, \frac{21}{28}, \frac{24}{32}, \frac{27}{36}, \frac{30}{40}$

The equivalent fractions of $\frac{5}{6}$ are: $\frac{10}{12}, \frac{15}{18}, \frac{20}{24}, \frac{25}{30}, \frac{30}{36}, \frac{35}{42}, \frac{40}{48}, \frac{45}{54}, \frac{50}{60}$

p34   **Challenge:**
The 12 possible decimal numbers are: 1·4, 1·6, 1·9, 4·1, 4·6, 4·9, 6·1, 6·4, 6·9, 9·1, 9·4, 9·6.

p35   A stands for 24·3   B stands for 24·8

p36   1st: Collins; 2nd: Wilson; 3rd: Brown; 4th: Jones;
5th: Smith

p37   The value of the third decimal place is thousandths.

p38   36 becomes 40   33 becomes 30   57 becomes 60
81 becomes 90   58 becomes 50   73 becomes 70

The carrot is about 14cm. The potatoes weigh about 2kg.

p39   **Challenge:**
Tottenham Hotspur v West Ham   34,000; 33,600
Chelsea v Manchester United   36,000; 36,300
Everton v Leeds United   35,000; 34,800
Arsenal v Sheffield Wednesday   27,000; 26,500

p40   Yellow: 40%   Red: 30%   Pink: 10%

p43   **Challenge:**
You can make 72 possible additions.